Mother Hubble's Cupboard

and Other Poems About Inner and Outer Space

by Sandra Lindow

July 2025

ISLAND OF WAK-WAK

Island of Wak-Wak
Västra Vintergatan 136
Örebro, Sweden
www.islandofwakwak.com
editor@islandofwakwak.com

Mother Hubble's Cupboard
and Other Poems About Inner and Outer Space

Cover design by F. J. Bergmann & Sandra Lindow
Cover art by the Hubble Telescope & NASA
Interior photographs by NASA.
www.nasa.gov

Typeset using KOMA-Script & LaTeX 2ε.
Printed and distributed via IngramSpark.
www.ingramspark.com

ISBN: 978-91-989599-3-2

Contents

Angels in the Architecture

(A)ngels in the architecture spinning in infinity

Paul Simon, "You Can Call Me Al," 1986.

There are angels in the architecture
of the universe, spinning the tapestries
of infinity, magnificent bodies
of making and unmaking,
sublime and ultimately unknowable,
pillars where stars die and are reborn.

If this is too much to comprehend,
take a walk on a 40-degree day in December,
see how the white mist rises from the lake,
melting the ice below,
know that the ice will come again,
but so will the warmth.

There are angels in the ice and in the water,
angels in the clouds that form above the surface,
angels in the grass that is one day lifeless
and the next day green,
angels in the crocus corms bursting
outward with expectation.

The architecture of the universe
is everything everywhere angels, their backs
spanning light years, their nebular arms linked,
their hair flowing unbound, their breath
flowing even into us — you can feel it, see it:
Take a breath, let it out, look around.

Defining Fire: The Physics of Poetry (and the Poetry of Physics)

Burning
like a cooling star,
it is a controlled reaction
set in action,
the weight of our dark matter
balanced by
the energy of our burn.

Gravity pulls us inward
toward an unspeakable,
unknown center,
but in falling,
we burn brighter
with heat
and light
of words.

Mother Hubble's Suite: In Homage to the Service of the Hubble Telescope

Mother Hubble's Last Will and Testament: June 13, 2021

Mother Hubble's cupboard is filled with stars
She remembers how her eyes mirrored anticipation
the first time she looked outward to see the universe,
its mysteries of birth and growth, as she captured

sustaining sunlight with her golden arms.
planet, galaxy, and star, stored like blushing peaches,
canned at the height of their succulence, rainbow revelations
so precise that we can almost taste their sweetness.

Now in the event of her demise, she sets heavy glasses aside
to bequeath a legacy of surprise and sparkling admonition:
Children of the light, you abide from darkness to darkness,
your fingers deep in the mane and marrow of your charger,

the living, breathing body you ride bounds light year
spirals into the future. Know that you cannot control
your own lives, your decisions based on incomplete
information, the dark matter of what you cannot see

ever working to defeat your most perfect plans.
Relax your grip and wonder in the vaulting of the sky
how the great blue beast you ride cries under the weight
of your self-indulgence. She's all you have, ride gently.

Mother Hubble Flies in Finery

With thanks to George Gordon, Lord Byron
and Chocolove "Almonds & Sea Salt"

> Hubble revealed that the Big Bang likely occurred exactly 13,799 billion years ago (plus or minus 20 million years!)
>
> Jim Bell, *Hubble Legacy*, p. 8

She flies in finery through the night,
acclaimed best-dressed in the starstruck skies:
high-healed photos, the best of dark and bright,
are captured through her camera eyes,
revealing to all the farthest light
that atmosphere to gaudy day denies.

Diffraction more and blurring less,
an unimpaired view of wondrous grace,
the jeweled night sky's raven tresses,
echoes of light across infinite space:
optical and ultraviolet composites
layer radio, x-ray, and infrared energies.

Her telescopic heart and mirrored brow
capture a history so eloquent,
mapping time by Big Bang's enduring glow;
myriad recordings of star lives spent
provide perspective for us down below
to seek peace and live love that is innocent.

Mother Hubble's Speeding Ticket: May 11–24, 2009

After Alberto Rios, "December Morning in the Desert"[1]

Her eyesight presbyopic, she must be refitted, **this**
grand dame photographer, known for her **great**
starstruck photos of faraway places, **hot**
bodies, cold spaces, butterfly wings, but her **rod**
has been speeding five miles per second, cosmic **skid**,
of rapturous road race glee across the continent **of**
North America, ten minutes flat out, and **the**
speeding ticket notes that she's been "dazzled by **Milky**
lights so lovely that her pedal hit metal all the **Way**."
White knight bodycams record a sunny grin **across**
her solar array and leave her beaming into the **asphalt**
vastness of an infinite star-spangled **night**.

[1]The last word of each line read combine to create a line from Rio's poem.

Mother Hubble on the Nature of Light

(With thanks to Charles Matson Lume)

Light is quicksilver, slippery,
nearly unmanageable;
I can honor but not keep.
It slips through my apron pockets
leaving only imprints of its passing:
traces, tracks, that can be followed,
wide-eyed, and found
beyond dust bunny clouds
through ultra-deep field sight,
back to the beginning of time,
an aching ember,
remembered in the Crab Nebula
a pulsar blooming at its heart,
thirty beats per second,
all that remains of a massive star,
living fast and hot,
dying for billions of years.

Mother Hubble's Happy Birthday Wish to Brother Webb, January 24, 2022

> The James Webb Space Telescope will find the first galaxies that formed the early universe and peer through dusty clouds to see stars forming planetary systems.
>
> (NASA)

Like Laika, Soviet sacrifice, sent to space without return,
your one-way mission to capture stars at birth.
Tall as three-story buildings, wide as tennis courts,
you were folded like an origami crane into the Ariane rocket:
a nested bird whose mirrored eyes were made
to measure the breadth of the universe.

When you were freed near a million miles from earth
your wide sun screen wings came first,
unfolding outward, shading the sun's hot burn,
then the honey-combed mirrors at your heart.
Like Laika, your life transfigures memory,
understanding becoming your birthright.

Mother Hubble Tames the Unicorn Black Hole

> Scientists have discovered what may be the smallest-known black hole in the Milky Way galaxy and the closest to our solar system — an object so curious that they nicknamed it "The Unicorn."
>
> (*Monthly Notices of the Royal Astronomical Society*, April 22, 2021).

Mother Hubble is not virginal:
She's lain with many a star:
her intercourse far more personal
than worshiping from afar.

Affairs well known in the galaxy
have bought her wisdom and experience.
She looks outward with acuity,
comprehension linked with patience.

Within Monoceros Constellation,
a black hole circles a red giant sun
attracting its lofty attention
without causing its demolition.

Born from the death throes
of a massive mother-father sun,
the Unicorn is small as black holes go:
a dark horn of despair, an orphan.

Seeing beyond its spell of invisibility
Mother earns the dark one's trust
She reaches out with curiosity,
not with exploitation or lust.

The Unicorn's tale is dark in its telling
of two who yearn but ever stay apart,
and Mother listens without judging,
holding their stories in her heart.

Mother Hubble's Elegy for RHESSI: (2002–2023), April 19, 2023

*The Reuven Ramaty High Energy Solar Spectroscopic Imager (*RHESSI*) was launched into orbit by the Pegasus* XL *Rocket on February 5, 2002. Before* RHESSI, *no gamma-ray images nor high-energy X-ray images had been taken of solar flares.*

Blind, my sister fell from bright into night
burning through atmosphere into the dawn,
charred loss on hard sand, hot Saharan sun,
she, once a dancer of unquenchable light,

Gamma Ray Girl spinning in low altitude flight,
dazzling as the day that she was first born,
Pegasus birthed to learn secrets of the sun,
capturing flares with intimate X-ray sight.

How can I mourn the loss of one such as she
whose pillow of sky I shared from afar,
who gauged lightning's electron velocity
with a heart-filled imaging spectrometer,
whose open eyes beheld blazing reality,
precisely mapping the shape of our star.

The Butterfly Effect

> The wings of the Butterfly Nebula are a 2-light year pair of high speed . . . high temperature . . . lobes of gas being jetted out in the death throes of one of the hottest stars in the galaxy.
>
> Jim Bell, *The Hubble Legacy*, p. 127

Mother Hubble mourns
the dance of death and beauty:
how thick layers of gas and dust
thrown outward by intense stellar wind
become illuminated
by the searing heat of loss,
ionized to form turbulent
knots, edges and walls.

Fierce and terrible,
as the hallowed hollowing
pain of grief,
but from a distance,
fragile as a Monarch's wings,
serene as a cathedral window,

perfect as that dazzling July day
nearly four months from your passing
when one landed on my hat
and chose to stay for a while.

Creation, Black Holes, and Big Bangs

Aristotle, and most of the other Greek philosophers, . . . did not like the idea of creation because it smacked to much of divine intervention.

Stephen J. Hawking, *A Brief History of Time*

God Smacked

(a golden shovel after Mark Doty)

Why do creationists sipping their coffee
think creation could be easy as bakery,
sun and stars molded like dough, dropped
and fired like bagels of enlightenment,
stones with dinosaur bones buried for mystery,
less celestial effort than breakfast, everything
alive, bacteria and bison, designed more perfect
than any dazzle of dizzying happenstance,
sea and sky separate, as blankets and sheets
lions and lambs predestined by teeth

Too Many

One, thinking windows to wonder,
wanted silicon but he-she was outvoted;
the bubbling pot in the heat
of the yellow sun stank
of carbon and ozone, thick and hot,
but Shalimari kept stirring resolutely
while Biblionni and Jeardana
argued over ring linkages,
a heated philosophical discussion that
distracted them from their jobs,
and Principinna lost the recipe
when the wind picked up, whisking
spirals in the brew,
Jeshurra and Ahoolin adding
elements willy nilly
when no one was looking.
Finally no one could say how
the clarity they'd first envisioned
had been lost to this rank, chaotic soup,
but some saw a certain poetry[1]
in the rough broth,
a self-sustaining eat and be eaten ethos
that was, nevertheless,
for celestial table,
totally inedible.

[1] *Biopoiesis* is a natural process by which life arises from simple organic compounds.

A Matter of Light

> We could be living inside a black hole and not even notice
>
> John Gribben, *Unveiling the Edge of Time*

Within the body of the dark mother,
the contractions continue,
all light turned inward.

Inside the dark fruit
of her intergalactic womb,
light pulses to light.

She is oblivious, our goddess,
starry-eyed romantic
for whom birthing is commonplace
and the internal issues
of small circulating bodies
mean nothing.
Let events take care of their own horizons.

She pops cosmic bon bons
between her limitless lips
and with great, languorous spaciousness
she yearns for the Other, the Spark
that cleaves her and opens her
and sets her heart on fire.

Black Hole Haberdashery

> The equations of general relativity also suggest that black holes have no 'hair,' or distinguishing characteristics aside from mass, electric charge and angular momentum.
>
> Emily Conover, "Gravitational waves offer new view of dynamic cosmos" *Science News*, Vol. 190, No. 13, 2016.

Black holes are bald,
and are uncomfortable about it.
Preferring to dress with timeless charm,
they'd wear bowlers like Steed,
if they could get them mail order
but they've got no hands
and their voices are ultrasonic,
so they swing space-time continuum
like lightless Jack-o-lanterns,
their black umbrella event horizons
swallowing everything in touch
in search of that vintage haberdasher
at the end of the universe, the promise
of quality and best Big Bang for their bucks.

Dreaming Black Holes

> The set of events from which
> it is not possible to escape.
>
> Stephen W. Hawking

In the dream
when I boiled, skinned and ate Schrödinger's kitten,
there was no uncertainty inside the lidded aluminum kettle;
I knew when the kitten died, a last look of reproach
as it was dropped into boiling water,
the once perky, pinkeared, tigerstriped baby
wantonly destroyed, frozen in memory's event horizon,
black holes that were its eyes.
"This is my body you eat."
It was as if I wanted to know what it took
to make something innocent die.

For I have seen black holes
in the eyes of abused children, and wondered
what star was lost by starvation, beating or incest,
when the future collapsed upon them
creating the naked singularity of innocence destroyed,
children trapped by past evil forever still about to happen.
Pleasure, self-esteem, love, trust, time
and the bright space of childhood
crushed by the black maw inside,
the membrane of memory itself worm-holed
in order to survive. I dream of them
reaching but never touching,
spun away by the churning silence.

Light years ago, you were energy,
dynamic radiation, dark force that drew me
across uncharted space
irrevocably as a star gazer's charm,
and you caught me, an event on your dream horizon
where in darkness I was compelled
to fall around you forever,
having touched your naked singularity
when I held you in my arms.

The Cold Dark Matter

> a cosmological constant to show how matter in an expanding universe dominated by cold dark matter could lead to the formation of great wall, great attractors and other huge aggregations of galaxies.
>
> *Science News*, 1/5/91

In her expanding universe
it is a cosmological constant,
the cold dark matter—
events long ago lost
when innocence exploded,
A Big Bang that threw her,
Staggering and unwary,
Into the blinding light;
Events that, though invisible,
Still shroud and shape the future,
buried secrets, impossible desires,
enormous unassailable walls;
the cold dark matter, the naked abomination,
shivering in the vacuum of inner space,
only a breath left to warm
the soul at Absolute Zero.
In her expanding universe
It is the foot on the brake,
it is the hand that . . .
the mouth that . . . , the words that . . . ,
the silence that . . .
it is the cold dark matter that
ravages the universe of
love.[2]

[2] "what is not remembered cannot be reported . . . it is not unlikely that more than half of all women are survivors of childhood sexual abuse trauma." (E. Sue Blume, *Secret Survivors*)

Chandrasekhar's Limit: Naming the Black Hole

Chandrasekhar's Limit: The greatest possible mass for a cooling star, above which it must become a black hole

Her night work took her to uncharted places,
Long hours, little monetary reward.
The lenses, though large and expensive,
Reflected nothing of the straining darkness
Behind her eyes and an internal singularity
It was becoming difficult to ignore.
Clutching a sweater across cooling shoulders,
Methodically searching the minutia
Of a vast, unwelcoming sky, she measured
The light of a faintly glowing body,
And found it strangely red shifted,
A pulsing radiation repeated by her racing blood.
Having exhausted all available energy,
A star above Chandrasekhar's limit had collapsed,
Unable to support itself against its own gravity,
A collapse of predictability
So bizarre it had to be hidden
Behind a dimly reflected event horizon.
Having pinpointed the location, she realized
She'd long sensed its attraction,
A set of events she couldn't escape
Even across light years.
Though not the first to observe the phenomena,
She could name it if she wished;
In a fading hand beside her initials
She wrote in the ledger
The exact location,
The time, "12:59:59 forever," then
"Light's Suicide".

Warping the Fabric of Space

> Moving at the speed of light, gravitational waves are traveling distortions of space and time, faint echoes of catastrophically violent events, including the big bang.
>
> Associated Press 3/21/93

Warping the watered silk of space,
moving at light speed,
gravitational waves
lap intergalactic shoals,
faint echoes of catastrophically violent events
whisper the Big Bang

Memory warps the watered silk
of neural space.
Traveling distortions of space and time
spread outward,
sweeping the cold, black silences,
lapping emotional shoals,
faint echoes of catastrophically violent events,
whisper what was once torn asunder,
whisper the Big Bang.

An Introduction to Alternate Universes: Theory and Practice

> Our Universe is just one among very many "bubble universes," all popping out of the general medium of the Big Bang like bubbles forming in a glass of beer. Somewhere perhaps there are many universes more or less like ours, some very similar to and others radically different from the universe we call "home."
>
> John G. Cramer "Other Universes II," *Analog* 11/84

I. Universal Soup

Inside the dark cauldron
universes bubble,
each bubble emerging,
breaking, becoming twinned —
"Double, double toil and trouble"
when life begins within.

II. Universal Tree

A time tree of universes
rooted in Yggdrasil,
formed by branching possibilities,
the crotch of each branch,
a crisis where time splits,
heralded by the crow, "What If?"

III. Dream Door

A microcosm of macro universes:
the shiny veneer,
wood of would,
the lintel is a shibboleth;
a handle the size of a sonnet
opens to a poet's breath.

IV. What If?

Somewhere onions dream,
whales sing oratorio; Elvis lives.
Universes not yet conceived
quicken behind a dreamer's eyes,
providing magic starship rides
into alternate possibilities,
but getting home is tricky.

V. The Glass

Is cold,
I take a sip.

Quantum Theory

In order to understand the Origin of the universe, we need to combine the General Theory of Relativity with quantum theory.

Stephen W. Hawking, "The Origin of the Universe"

Cosmic Requiem

> It is a tone no one can hear, generated by a body no one can see, yet its reverberations may be huge.
>
> R. Cowen, "A Low Note in the Cosmos," *Science News*, Vol. 164, #11, 9/16/03

I.

Far across the night
a massive star collapses
swallowing its light.

Light implodes on light
as gravity commands it;
a shroud folds into night.

No escape for light;
a new black hole demands it;
a grave enfolds the night.

The center's infinite density
confuses relativity,
crushes cosmic secrets,
breaking space and time.

II.

A silence at the center
swallows all that enters
but a dirge transcends the core.

Light transfigures to sound;
concentric sound waves spiral;
midnight ripples the pond.

One note, B-flat, resounds,
fifty-seven octaves down,
largo, rallentando, decrescendo.

From the hub of the Perseus galaxy
the lowest note in eternity

resonates forever into night;
requiem aeternum, perpetual light.

Reflections on a Geode in the Hand

> Matter is nothing but gravitationally trapped light.
>
> Matthew Fox

1. Mass

The stone
has a cathedral in its bones,
flying buttresses and domes,
a place of arches and choirs
where electrons fly
and do not fly alone.

2. Energy

The stone
has a tavern in its soul,
a place of warmth and light
where protons and neutrons
cluster a nuclear hearth
singing old songs,
drinking the convivial
quarks of poems.

3. Relatively Speaking

The importance
of one stone in a plowed field:
a turning of time,
unifying the lit lens
of an event horizon —
Ein Stein versus Einstein

4. Time

Rocking ragged
in a landscape of stone,
tucked in by sand,
the old farm sleeps soft
in the cradle of its bones,

the rust and ravel of change
stilled for a sedimentary second,
taken for granite,
but change remains:
gravitons and graves,
bosons and bisons,
and old as the hills
with charm and strange
the earth awakes
for another day.

5. Unified Field Theory

A stone
found in a field
explains everything.

Studying String Theory

. . . a colossal number of strings all undergoing the same orderly, graviton pattern of vibration . . . an enormous, organized array of similarly vibrating strings is known as a coherent state of strings.

Brian Green, *The Elegant Universe* pp. 377–378.

Universal violin, ultra-thin
Stradivarian skin,
turning out, turning in,
turning back where it begins,
Moebius-waisted midnight velvet,
the music of gravitation,
graviton vibration,

plays an orderly coherent
vibrational dance, a subatomic
timeless, resonant romance,
transforming cosmic fabric
so fine as to be invisible within
visible light's space-time design.

Thunder rolls,
an orchestral sound.
plucks EADG from a bow
of boson clouds
and the universe sings,
"What now? What if?"

Finding the God Particle

> Without . . . God particles, atoms would have no integrity, so there would be no chemical bonding, no stable structures — no liquids or solids — and, of course no physicists and no reporters.
>
> Stephen Fried, "The Race for the Secret of the Universe," *Parade*, 2009

It is something and nothing,
the God particle;
everything else is mass:
tables, chairs and books,
the lint we remove from dryer filters,
even the air we breathe.

But the God particle
is like the matrix of a poem,
invisible force,
slowing what is born
massless at light speed,
pulling, binding the words together
until it takes form.

We cannot see where it begins
only the words of where it's been:
the sub-atomics of creation,
colliding in darkness,
taking shape in the ashes
of beauty, desire and pain.

At Fermi Lab
scientists find the secret
of universal cohesion,
not in their particle accelerator:
but in their luminous dreams.

Goddess Principle: The Proof Is Pie

The named is the mother of creation.

Lao Tzu, *Tao Te Ching*,
John H. McDonald, trans.

The observable universe can be thought of as a sphere that extends outwards from any observation point for 46.5 billion light years.

If the universe is round,
it is more goddess that god,
a triumph of feminist principles
repeated in quantum composition,

like a curving casserole of comfort food,
carefully constructed
and considering the uncertainty principle,
never measured exactly, acting
on intuition as good cooks do,

lasagna layers of subatomic particles
lovingly laid: bosons, muons,
leptons, gluons, all snugged together
like Aunt Elvira's Celery, Chicken and Rice,
Corn Chex Delight

a quarky combination of flavors:
up, down, bottom, top, charm and strange,
each scoop containing all,
but not in any particular order,
pions, the lightest mesons and hadrons,
supporting nuclear recipes since
the ultimate oven beginning of time,
reflective of Bisquick quiche pies,
the transformative nature
of matter and energy as that sticky
unpalatable slop of egg and cheese
combines, expands: busy quickness
of particles in motion, rising
hot at four hundred degrees—

or fifteen million degrees,

the golden pies resulting, fluffy
transcendent, luminous,
absolute proof of a balanced
and eminently sensible goddess

Quantum News Flash

Higgs Boson at last
nabbed as it passed
Schrödinger's cat

Flirting with Dark Matter

> To find evidence of the Higgs boson flirting with dark matter, scientists must learn how to see the invisible.
>
> "The Secret Life of Higgs Bosons,"
> *Symmetry Magazine*, 02/20/18

Feeling
most invisible
she met him
who was Dark Matter
and pressed against his impossible,
discovered convolutions,
an intercourse
of light

Creation: Dark Matter Dating App

> The Higgs boson has existed since the earliest moments of our universe.
>
> *Symmetry Magazine*, 02/20/18

She was a bubbly Higgs Boson
from somewhere and nowhere
up north of a Venus Beach
that hadn't happened yet.
What she wanted
was hard to pin down in her profile.
She was all over the place
in the qualities she sought.
He was Dark Matter when she met him,
all Camp Fire and Camel cigarette.
At first, she wasn't there for him,
but then she was,
and when they came together,
the expanding universe spun.

Designated Responsibilities for Spokesparticles

> (T)he Higgs boson, the spokesparticle of the Higgs field, must have some relationship with dark matter particles.
>
> *Symmetry Magazine*, 02/20/18

> Thirty spokes / meet in the hub.
> Where the wheel isn't / is where it's useful.
>
> Lao Tzu, *Tao Te Ching*, Ursula K. Le Guin, trans.

Coming to you live from Higgs Field
Today we interview Higgs Boson, Spokesparticle
for Universe Corporation,
regarding her designated responsibilities
for particle outreach interactions.
We are told that "spoke" refers
to the support structure of a wheel
as well as officially sanctioned speech.
Higgs's ansible communications
describe mundane and mystical activities
"wheeling and dealing for galactic unity,"
suggesting intriguing questions for future study
by physicists and theologians.
The quantum level excitation of the crowd
is evident outside CERN Hadron Collider
where Higgs reportedly meets
with Dark Matter Industries,
a cognitively seductive interchange of ideas.
Rumors of self-coupling may be unfounded.
And now here's Higgs [applause] . . .
Well, it looks like we missed her again,
but this busy woman has left us proof
of her continuous positive regard:
twin neonatal tau fermions [applause]
Will someone please call the nanny?

Coming Out of the Dark

Darkness born from darkness
the beginning of all understanding
being and non-being produce each other

Lao Tzu, *Tao Te Ching*, John H. McDonald, trans.

Come on, come on, do the locomotion with me.

Carole King, Gerry Goffin, "Loco-Motion"

Forests and farmlands slipping into darkness,
ten miles west of the poetry workshop,
three deer suddenly dance before me:
"Come on, come on, do the loco local motion
with me." A rainy mist-swept two-lane,
fortunately, I'm alert, can stop
short of their heedless winter worn bodies;
thanks be, no one tail-gating behind me,
but my case on the rear seat is tossed,
spilling poetry, that quirky, quarky
quantum matter of bright and dark,
spirit and memory, and I think
how dark matter dances unseen
across apparent emptiness, telescopes
like headlights sweeping parsecs,
only unexpected movement in the eyes of galaxies,
reflecting an undetected Otherness,
Higgs bosons and dark matter interchanging,
creation dependent on decay,
expansive starlings fading to doves,
the constant reformation of matter,
discovered in the night sky,
deer dancing on a dark matter highway,
tessellations of Escher art.

Sex and the Theory of Everything

> Why does the universe go to the bother of existing?
>
> Steven W. Hawking, *A Brief History of Time*

> After three-hundred-year sex, anybody can fly!
>
> Ursula K. Le Guin, *The Telling*, 52–53

In the bright bath of hot August sunshine,
Japanese beetles mate on a bean leaf,
their iridescent wings refracting light,
an oily shading of night sky colors:
black becoming indigo, teal, and rose,
bodies locked in silent intercourse—
nearly motionless Tantric connection,
a weaving that's fractally reflected
in perpetual cosmic coitus,
infinite orgasm, light and dark matter
everywhere enacting acts of becoming,
tiny explosions of new information,
and then from quantum to pollen to person,
mountains, oceans embrace, and it all fits.

Space and Time

> . . . when one combines general relativity with the uncertainty principle of quantum mechanics, it is possible for both space and time to be finite without any edge or boundaries.
>
> The progress of the human race in understanding the universe has established a small corner of order in an increasingly disordered universe.
>
> Stephen W. Hawking, *A Brief History of Time*

Dwarves

After lunching on hydrogen and helium,
the seven dwarf stars, knowing the bondage
of carbon and oxygen, collapse heavily
into their dark matter beds and dream
of a gentle snow white space that sings
them the songs of their ice diamond souls.

Leaps to the Moon

Above and beyond the rotting
watermelon rind and banana peels
in the mulch bin,
Venus and Jupiter closely
flank a brilliant crescent moon,
a configuration that won't
be seen again for another sixty years.
I will be mulch for moonbeams then
so for now I stand coatless,
transfixed by bannered purple-teal sky
while the breath of December
fills me to my bones.

Sonata for the Moon on the 120th Anniversary of Georgia O'Keeffe's Birth[1]

Since I cannot sing, I paint.

Georgia O'Keeffe, 1922

Andante. Beyond the comforting
bulk of the house, November sky
shades from azure blue to piano black.
The wind blows cobalt cold,
an arpeggio of dry leaves,
and a fingernail moon lingers
behind a ripped gauze of glissando cloud.

It is a picture O'Keeffe could have painted
if she who craved light
had not left Wisconsin. Perhaps
this merest curve of ivory moon
would become the hint of breast
on which the night nuzzles to suck.
Secret night moon woman,
baring her breast, because
it is what she does though she knows
it will never be enough
to keep winter from the womb
of the world. O'Keeffe
would add a few birds.

[1]Georgia O'Keeffe born November 15, 1887

Seventy-Year Moon:

After Issa[2]

Brilliant stars and a seventy year
moon, closer than seen in a lifetime,
is an intimate song floating above rooftops;
it sings descant to bare boughs of cottonwood,
true notes of an ancient promise
that the galaxy goes on beyond human catastrophe.
You, who stand in the shadows, come look:
Too much has happened to look away, you
must seek songs in yourselves or truth will
pass like battalions goose stepping
in full dress parade, leaving the powerless, only
a scrap of confetti and a bone of contention;
hurry, for the first time, see.

[2]The first word of each line forms a haiku.

Solstice Moment

after Basho[3]

Pale water sky the first day of *winter*,
a season of darkness, solstice *seclusion*,
the longest noon shadows *sitting*
on the front yard like spectators, *propped*
and waiting as anorexic sunshine races *against*
slushy streets, arthritic maples, and *the*
rusty old street sweeper trundles by, *same*
old earth, war and bigotry threadbare *worn*,
while a minute more sunlight sprints from the *post*.

[3]The last words of each line combine to make a Basho haiku.

Moondome Drag Show Review[4]

Floating negligees
Insouciant layers of pink and pearl
Miners mine for laughs
Hefty harvest of hairy legs
Setting earth sings backup blues

[4]The gravity of the moon is 1.6249 meters per second squared, or 16.6 percent that of Earth. A person who weighs 150 pounds on Earth weighs just 25 pounds on the moon. The moon represents the female principle, the Goddess's mirror reflecting everything in the world.

Intragalactic Season

Debutant Saturn
engaged seven times
kept all her rings
flaunted them
to the high
heavens

bottle trees on Mars

red silicon most common
black loam humus none
the first greenhouse hops harvest
empties growing bottle trees

(A farm girl writes home)

The Venus of Letting Go

Aphrodite, Innana, Ishtar,
sister planet, morning and evening star,
who proves tawny template for letting go:
oceans vaporized, oxygen, hydrogen
photo-dissociated and swept outward
by solar winds, a wedding with the night:
Like the walk to the woods on the family farm,
the last Green Contenders grown in the yard,
the jingling keys to the little blue car,
so much to lose and still keep shining bright
amidst the volcanic rubble left of life,
she descends to dark, and returns as light.
She glows, an ode to nothing left but love
throughout the long, dark star struck night.

USS Cybership: The Restless Night

The universe is the mind of God.

George Berkeley

The Night is always going home
to a place that slips forever farther away—
each planet, each star, each galaxy less familiar,
at first subtly changed, a jump-point uncharted,
a satellite misplaced, then odder and odder
until not even the universe is the same,
for this is now an alien place
where novae bloom like bad ideas:
surreal cabbage roses in zirconium sill boxes,
huge and decadent, their heavy emanations
drugging twilight sleep, and nowhere to dock,
no familiar berth, the number on the gate
synonymous with home, no place to rest
at the end of a solar day, but, finally,
when all memories, like gantries have fallen away,
only the cool, gray passages of a brain where she wanders
like a patient forgotten when the madhouse moved,
restless heavy with fatigue, unable to stop looking
for a berth strong enough to hold her, umbilicals
to ease her, and a place to rest her feet.

Cyber-Islands

That summer we caught rainbows
in the hydroponic pools,
voracious, silver shimmers,
a child's hand wide,
who had a taste for synth-ham sandwiches
and standard ration slumgullion stew

We eased them up between the slats
of the service walkways. Too small to eat,
they were food for the spirit,
and we threw them back, rainbows falling,
until the alterday shift came on duty,
dome lights dimmed,
and we agreed to fish no more.

We were first on Alpha Station born,
minnows then, rainbow wigglers,
net-leaping hackers spinning joy
through a screen of dusky green,
endlessly summer days.

Now we are slower, our rainbows subdued.
Initialized, we move languidly
away from the computer docks,
swim eastward through dark silk,
toward green islands,
once immeasurably far, now near.

There we will sink into tangled cradles
and dream of channels, deep and cold,
that suddenly widen and dazzle,
a dancing of starlight in our eyes
that seems to last forever.

Yolanda's Research

an abecedarian

Zeno's Paradox suggests true success impossible,
Yet yellow-eyed Yolanda, the youngest
Xenologist at Xenobia University
Wanted to establish alien first contact when
Virtually all university illuminati concluded impossibility,
Urging intensive research on organic traces
That had been gathered by Rosetta's Philae lander,
Star stuff sniffed out of a comet's atmosphere.
Really not ready for the microscopic life, Yolanda
Quit her quantum study quietly,
Putting her energy into classroom performance,
Only to be broadsided by a bombshell:
Noting advances, she'd had her genome tested—
Mucus from her mouth sent to a Neo Mormon
Laboratory where results left her lightheaded,
Keeping her up at night for weeks.
Just under ten percent of her DNA wasn't human!
It ruled out Neanderthal and Denisovan.
Helen of Troy might have been her mitochondrial mother,
Greeks, then Spaniards, but what else could have
Filled chromosomal coffers, launching genetic ships?
Evidence suggested organic compounds
Disturbingly similar to the lander's findings—
Could a comet be key? She was afraid to believe,
But each step halved reached the same conclusion:
Alien legacy discovered—star stuff in herself.

Virgo, of Stars and Bread: An Exquisite Corpse

From birth we are like bread.
The cold fact is that
Stars appear to twinkle
people die
the yeast on which we rise,
temperature variations
too soon or too late
create turbulence
heat activated
in earth's atmosphere
we eat and are eaten—
carbon-chained flesh,
the sugar consumed
with loose threads and fragments
experience, the salt that holds the leaven,
changes the refractive index
life's agenda uncompleted
by milk's sweet innocence uplifted,
firing and fading like stars,
we lighten the Milky Way.

Perspective: Sunset Through the Telescope

Looking westward from the patio rooftop of the hotel,
the sky is pale linen warming to luminous yellow voile
above the seam line of a boulevard and a lacy selvage
of green deciduous trees, the fabric of late July twilight.
Walking past tables and lounge chairs, I wonder
how the sky weaves its ever-changing colors,
a felicitous twining of molecules and light.
At the edge of the building, a thin, bright-eyed man
is happy to show me his prized reflecting telescope:
The metal sleeve of black enamel and silver edging
has a microscope-like eye piece attached along its side.
Inside is an array of mirrors, he says. Cautiously, I bend
to look through the lens and am startled to see the last rays
of the setting sun vividly revealed at a ninety-degree angle,
flattened as if I viewed film running through a projector,
a vision that excludes and estranges, forcing me to reconsider
perspective, and how human vision clothes the world,
providing the alluring but dangerous optical illusion
that all revolves around us.

the setting sun tucks in
sideways behind the trees
space station earth

Matreshka: Cracking Universal Mystery

Picture then the universe
In its awesome entirety
Curving like a Russian Matreshka doll,
Voluptuous, eggshaped, female,
Starryeyed, bright,
Handpainted to a perfect gloss,
Hiding pregnant purpose
Beneath her spacious apron.

Around her applecheeked,
Peasant's head, her babushka
Shades evenly toward doppler red,
Predicting the ageless secret
Of cosmic distance.
Her shawl contains the galaxies,
Knotted, forever falling
From her shoulders.

The hem of her multitudinously
Starflowered skirt
Firmly encloses the inner darkness.
Hem flows unbroken to hem
And within her inner space
Another doll curves
Another and another,
Each as perfect and intricately formed
As the First
Until in some final subdoll region,
A seductive, unmeasurable
"Tendency toward" Matreshka
Exists, though we cannot see her.

As we crack each half, we think
Can the hand, the mind
That cracks the halves create
Through power of expectation
The shape that nests within?

Galaxy, solar system, planet, continent . . .
Woman and the child that blooms within,
A narrowing of size,
A widening of inner spatial distance,
Cells within tissues,
Mitochondria within cells,
Molecules, atoms, ions . . .
Gluons, quarks, . . .

Where does life begin? Where end?
As form nests within form
And all are enclosed
Within the body
Of some singularly female
And infinitely pregnant God.

Like those who sought the Grail,
We cast aside our uncertainty
And set forth across the universe.
An arc the shortest distance
Between two points,
We bound light years in a blue
And silversailed Matreshka,
Scanning the voluminous dark,
Swinging the gravity wells
Of goldenspangled stars,
We learn only after eons spent
That now in our ancient,
Seachanged innocence,
We can see the chalice
Of our journey's end,
Curved as the open cup of a Matreshka,
Is back where we'd begun.

Mother Hubble's Cupboard: Notes and Acknowledgments

Hubble Telescope

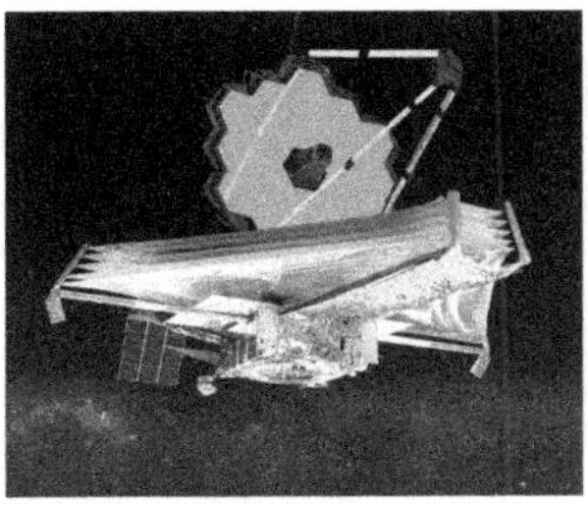

Webb Telescope

Rhessi

Notes on the Poems:

"Angels in the Architecture" Inspired by the Webb Telescope's photograph of the Pillars of Creation. "Creation" is an apt term to describe the environment in this part of the galaxy. The dark molecular clouds contain the remains of previous generations of older stars, which are the building blocks of new stars" (Bell, 104).

Mother Hubble's Suite:

"Mother Hubble's Last Will and Testament" On June 13, 2021, the Hubble sent a message that it was shutting itself down due to a problem in its payload computer. It took a month for scientists to develop a solution for the problem. Although the Hubble is once again capturing the universe, the end is near. It is expected to be taken out of orbit sometime this decade.

"Mother Hubble's Speeding Ticket" This is partially based on a speeding ticket that I almost got; however, when reading Jim Bell's *The Hubble Legacy: 30 Years of Discoveries and Images*, I was impressed by the difficulty of docking with the Hubble and servicing her because she is moving so fast.

"Mother Hubble Flies in Finery" Despite the high-healed pun, Hubble's layering (i.e. healing) of six different camera images is serious business. Colors from energy that the human eye cannot perceive are layered and given a "false color" composite that can be seen and appreciated. This gives us a better understanding of the nature of the universe. (Bell, p. 8)

"Mother Hubble on the Nature of Light" Inspired by an article on artist Charles Matson Lume "UW-Stout professor makes art with light" that appeared in the *Dunn County News*, Aug. 31, 2022. "Light is quicksilver. It is so slippery and nearly unmanageable. Light is also ephemeral, like us. Both are immortal for a limited time" (p. A2).

"Elegy fot Rhessi" The Reuven Ramaty High Energy Solar Spectroscopic Imager (RHESSI) was originally called a High Energy Solar Spectroscopic Imager or HESSI and also Explorer 81. Its primary mission was to explore the physics of particle acceleration and energy releases.

"The Butterfly Effect" The Butterfly Nebula is about 3,400 light years from earth. Its fiery gas jets wings have a temperature of about 36,000 degrees Fahrenheit and are moving at more than 600,000 mph.

Creation, Black Holes, and Big Bangs:

"God Smacked" Gobsmacked is British slang for being too surprised to speak, but "God Smacked" refers to the theological obstetrics of creationism. In *The Woman's Encyclopedia of Myths and Secrets*, Marina Warner explains that Christian scholars such as Chateaubriand, believed that the earth was literally created in seven days and tried very hard to refute the discovery of the bones of extinct species, suggesting God consciously created "the appearance of pre-existence in order to test their faith" even going to the ludicrous extent that the fossils were deceptions formed of "lapidific juice" or "seminal air" (186). According to Stephen J. Gould, creationists argue that evolution is an "unproved and unprovable charade—secular religion masquerading as science." (*The Dinosaur in the Haystack*, 409). Also, I once worked in a restaurant that made doughnuts. Thus, the poem explores two visions of creation: God as doughnut maker of the universe and God as obstetrician, smacking new born life on the bottom so that it takes a breath and cries.

"Too Many" Biopoiesis is the natural process by which life arises from simple organic compounds, but the etymological and ontological parallels to poetry are obvious. How many angels can dance on the head of a Higgs Boson? An infinite number since angels are incorporeal; however too many cooks can still spoil the broth.

"An Introduction to Alternate Universes: Theory and Practice" There are many definitions of speculative poetry, none of them entirely successful. This poem works to define specpo as an intersection of alternate universes that are comprised by branching possibilities seen in the bubbles on a beer head as well as Yggdrasil. Yggdrasil is the World Tree depicted in Norse Mythology. The Norse God, Odin, sometimes called the Raven God, spent nine days hanging from it in search of wisdom. A shibboleth is a word or catch phrase used by a group that becomes an identifier for members of the group. In speculative poetry circles "What if?" becomes such an identifier.

"A Matter of Light" The Egyptian Goddess Nut (Sky) was believed to Swallow the sun at night. It would then travel through her body and she would give birth to it at dawn. In *Womanguides*, Rosemary Radford Ruether describes the body of the Goddess "encompassing the matrix of the universe, radiating the sun powers from her womb, the moon powers from her head" (36). A painted relief of this can be seen at the Temple of Hathar, Dendera, Egypt. The Babylonian Goddess Tiamat first gave birth to herself, Mummu, which according to Barbara G. Walker (*The Woman's Guide to Myths and Secrets*, 998) can be translated as "mother" or "churning." The ancient notion was that the solid earth was made by churning primordial fluid the same way butter is churned from milk. Here the concept is updated to include the cosmological equivalent of Godiva chocolates.

"Black Hole Haberdashery" John Steed, played by Patrick Macnee, appears in the popular British TV espionage series from 1961-1969. One of Steed's trademarks was his impeccable dress: usually a classic black suit, black bowler hat, and ever-present, sword-equipped umbrella, which could be used as a very effective weapon. A 1968 episode shows him in a startlingly white bowler instead of a black one.

"Dreaming Black Holes" Theoretically, the center of a black hole contains a gravitational singularity, a one-dimensional point where a huge mass is reduced to an infinitely small space, density and gravity becoming infinite and space-time curving infinitely. The laws of physics as we know them cease to operate. The Nasa Science website describes black holes as "a great amount of matter packed into a very small area—think of a star ten times more massive than the Sun squeezed into a sphere approximately the diameter of New York City. The result is a gravitational field so strong that nothing, not even light, can escape." Correspondingly, psychological repression can shrink those traumatic memories to a very small part of consciousness, but this only increases trauma's emotional weight until the light of joy is diminished if not extinguished entirely.

"Chandrasekhar's Limit: Naming the Black Hole" "If the total mass of the star is large enough (about three times the mass of the Sun), it can be proven theoretically that no force can keep the star from collapsing under the influence of gravity. However, as the star collapses, a strange thing occurs. As the surface of the star nears an imaginary surface called the 'event horizon,' time on the star slows relative to the time kept by observers far away. When the surface reaches the event horizon, time stands still, and the star can collapse no more—it is a frozen collapsing object" (*Nasa Science* Website)

"The Cold Dark Matter" Observations indicate that approximately 85% of the matter in the universe is dark matter, with only a small fraction being the ordinary baryonic matter that composes stars, planets, and living organisms. Psychologically, brains retain negative and traumatic material (or dark matter) much more readily than positive events. Negative experiences move immediately into long term memory while positive memories must be savored for them to be remembered long-term. Many abuse survivors lose the ability to remember positive interactions. The focus of therapy then is to help them rebuild their positive memory-retaining structures through bodywork, massage, and meditation.

Quantum Theory:

"Studying String Theory" One theory for the shape of the universe is that it would loop back on itself like a Klein bottle or Moebius strip. GDAE are the strings on a violin.

"Cosmic Requiem" According to an article on the Nasa website, "Sounds of the Ancient Universe," the sounds of the ancient universe can be "heard" in a sound file via the Planck Space Telescope. The article explains that "Before there were any stars or galaxies, 13.8 billion years ago, our universe was just a ball of hot plasma — a mixture of electrons, protons, and light. Sound waves shook this infant universe, triggered by minute, or 'quantum,' fluctuations happening just moments after the big bang that created our universe." The resulting ancient music sounds a bit like the whooshing of a street sweeper as it passes by on a bright spring day.

"Reflections on a Geode in the Hand" A graviton is a hypothetical quantum of gravitational energy, regarded as a particle

"Finding the God Particle" The Higgs Boson or top quark was proposed by Peter Higgs in 1964. July 4, 2012, its existence was confirmed at the CERN Large Hadron Collider in Switzerland. The CERN Collider is the world's largest and highest-energy particle accelerator. Built from 1998 to 2008, CERN has a circumference of nearly seventeen miles.

"The Goddess Principle: Proof is the Pie" Early creation myths generally present a symbolic gynecological view of creation, but this myth concerns cooking. The temperature of the interior of the sun is calculated to be 15 million degrees Celsius.

"Quantum News Flash" Schrödinger's cat has become an SF trope, taking on a reality hardly envisioned by Nobel prize winner, Erwin Schrödinger, when in 1926, he published this thought experiment to critique the Copenhagen interpretation of quantum mechanics. Schrödinger's cat-in-the-box paradox theorized that the cat would exist in a state of being neither alive nor dead until the box was opened. However, considering the cat's continuing popularity ninety-three years later, it can be argued that the cat will always be alive and lurking in the spaceways.

"Flirting with Dark Matter" If the physicists are right, the infinite flirtation between light and dark matter forms the fabric of the universe. Like the Way of the Tao, the mating of light and dark matter remains an unobservable, indescribable act: The Way that can be described is not the eternal Way.

> "Heaven and earth begin in the unnamed:
> name's the mother of the ten thousand things."
> Lao Tzu *Tao Te Ching* (Le Guin, trans.)

"Designated Responsibilities for Spokesparticles" My idea here is that dark matter and Higgs bosons interact as parents. "It could be that dark matter aids in the production of Higgs bosons, or that Higgs bosons can transform into dark matter particles as they decay," Darin Acosta, a physicist at the University of Florida. ("The Secret life of Higgs Bosons," *Symmetry Magazine*, 02/20/18)

The parallels between Taoism and quantum physics have led to considerable literary discussion. Two early books are Fritjof Capra's *Tao of Physics* (1975) and Gary Zukav's *Dancing Wu-Li Masters* (1979). "Wu Li" is the Chinese phrase for physics. It means "patterns of organic energy," but it also means "nonsense," "my way," "I clutch my ideas," and "enlightenment."

"Coming out of the Dark" The last three lines of "Loco-Motion" are: "There's never been a dance that's so easy to do. It even makes you happy when you're feeling blue. So come on, come on, do the Loco-motion with me." Carole King and Gerry Goffin wrote words and lyrics and Little Eva made it a hit in 1962. I remember from watching Dick Clark's *American Bandstand* that the Loco-Motion was usually performed as a line dance. Thus, when a line of deer suddenly appeared out of the dark, I thought of local *loco* motion, or crazy behavior, but then considered the constant interchange of matter and energy.

"Sex and the Theory of Everything" Einstein called it a unified field theory. The theory of everything (TOE), refers to the ultimate or master theory that is a hypothetical single, all-encompassing, coherent theoretical framework of physics that fully explains and links together all physical aspects of the universe.

In Indian philosophy, Tantrism teaches sexual techniques that promote divine union with the sacred feminine through prolonged sexual connection.

Space and Time:

"Dwarves" won 2nd prize in an SFPA Dwarf Star contest. I was influenced by the Disney's cartoon *Snow White.*

"Sonata for the Moon . . . " Georgia O'Keefe was born on November 15, 1887, in Sun Prairie, Wisconsin and studied at the Art Institute of Chicago and the Art Students League in New York. She married art dealer and photographer Alfred Stieglitz and lived with him in New York until 1929 when she started spending part of each year in New Mexico.

"Seventy Year Moon" When the moon is closest to us, its proximity makes it appear 14 percent larger and 30 percent brighter in the sky than an apogee full moon (Apogee is when the moon is furthest from Earth in its orbit). Therefore, on November 14, 2016, the moon was nearer to our planet that it had been at in seventy years and metaphorically illuminated concerns about the 2016 election and increasing far right politics in the United States

"USS Cybership: The Restless Night" I began reading the Heinlein YA books when I was in elementary school and immediately became hooked on the original NBC *Star Trek* series in 1966. Bishop George Berkeley was born March 12, 1685 and died January 14, 1753. He was known for his advancement of a theory he called "immaterialism" which contends that familiar objects like tables and chairs are only ideas in the minds of observers.

"The Venus of Letting Go" "Green Contenders" are a variety of green bean that grows well in Wisconsin. Along with the goddess allusions and the scientific depiction of Venus, this poem parallels the Sanskrit Mahamrityunjay Mantra or death-conquering mantra which celebrates letting go of physical attachments and the importance of change. The translation explains that the gardener must separate the "cucumber" from the vine for it to be of any use. The mantra is chanted for healing and rejuvenation.

"Moondome Drag Show Review" The gravity of the moon is 1.6249 meters per second squared, or 16.6 percent that of Earth. A person who weighs 150 pounds on Earth weighs just 25 pounds on the moon.The moon represents the female principle, the Goddess's mirror reflecting everything in the world.

"Bottle Trees on Mars" The first artwork of the first colony on Mars might be a variation on a bottle because there would be no trees to look at outside the dome. It is also unlikely that humans, even scientific humans, will ever entirely outgrow superstition. Glass bottle trees originated in ninth century Kongo during a period when superstitious Central African people believed that a genii or imp could be captured in a bottle. Later, colorful bottles were believed to attract and hold evil spirits when they were hung on trees. This practice was taken to Europe and North America by African slaves.

"Yolanda's Research" November 12, 2014, Rosetta's Philae lander touched down on comet 67P/Churyumov–Gerasimenko.

"Virgo, of Stars and Bread: An Exquisite Corpse" Virgo is the constellation of the harvest maiden. An exquisite corpse is a poetic form that combines lines from several sources. The above combines quotes from Dr. Edwin Shneidman's *Voices of Death* and *Science News* (2/25/89) with a connecting bridge.

"Matreshka: Cracking Universal Mystery" "Matreshka" is the first poem written in this series. It was begun around 1987 and published in 1989. Einstein's original field equations suggested that the topology of space could fold back on itself, leading to considerable discussion regarding the shape of the universe. Recent work in cosmology is moving away from this possibility; however, the psychological aspects of journey, change, and return are as valid today as when I began exploring the concept, as is the nature of the universal feminine.

Author Bio

Sandra Lindow grew up during the space race and has had a lifelong interest in speculative poetry. Her first poem, which was posted on the wall of her fifth-grade classroom began "Many years have come and passed from Pilgrim ships to rockets blast." *Mother Hubble's Cupboard* is her tenth collection of poetry.

Presently she serves as the Wisconsin Fellowship of Poets' longest serving regional vice president and SFPA's SpecPo Review blog's coeditor. She lives on a hilltop in Menomonie, Wisconsin. More can be found out about her at `https://www.wfop.org/member-pages-1-z#/sandra-lindow`.

Acknowledgments

"Black Hole Haberdashery," *Star*line* 41.1, Jan. 2018.
"Bottle Trees on Mars," *Star*line*, March 2018.
"The Butterfly Effect," *ETTT* 49, July 2023.
"Coming out of the Dark," *Eye to the Telescope*, 29, July 2018.
"Cosmic Requiem," *Magazine of Speculative Poetry*, Vol. 7 #1, Winter 2004–2005.
"Creation: The Dark Matter Dating App," *Asimov's*, July/Aug. 2019.
"Cyber-Islands," *Kaleidoscope*, 34, January 1997.
"Designated Responsibilities for Spokesparticles," *Star*line* 41.4, Fall 2018.
"Dreaming Black Holes," in *Strange Horizons*, 9/24/2001.
"Dwarves," *Magazine of Speculative Poetry*, Spring 2006.
"Finding the God Particle," *Strange Horizons*, 2009.
"The Goddess Principle," *Moss Piglet*, Sept. 2021.
"God Smacked," *Museletter*, Fall 2016.
"An Introduction to Alternate Universes: Theory and Practice," *Gyroscope Review*, 16/1, Jan. 2016.
"Matreshka: Cracking Universal Mystery," *Amazing Stories*, May 1989.
"Moondome Drag Show Review," *Scifaikuest*, Aug. 2018.
"Mother Hubble's Speeding Ticket," *Star*line* 45.4, Fall 2022.
"Mother Hubble's Suite," (five poems): "Mother Hubble's Last Will and Testament: June 13, 2021," "Mother Hubble Flies in Finery," "Mother Hubble on the Nature of Light," and "The Butterfly Effect." *Ionosphere* Issue 1, Jan. 1, 2024,
"Mother Hubble Tames the Black Hole Unicorn," *Dreams and Nightmares* #124, May 2023.
"Quantum News Flash," an earlier version of this poem was published as "At Last" *SFPA Contest page*, November 2018, `www.sfpoetry.com/contests.html`.
"Seventy Year Moon," *Leaves of Peace Anthology*, June 2020.
"Sex and the Theory of Everything," *Dreams & Nightmares*, 111, Jan. 2019.
"Solstice Moment," *Sky Island Journal*, Fall 2017.
"Sonata for the Moon on the 120th Anniversary of Georgia O'Keeffe's Birth," *Bramble*, Spring 2018.
"Too Many," *Dreams & Nightmares* 100, Jan. 2015.
"USS Cybership: The Restless Night," *Star*line* 22.2, March/April 1999.
"The Venus of Letting Go," *Dreams & Nightmares*, 105, Jan. 2017.
"Virgo, of Stars and Bread," `www.astropoetica.com`, 2003.
"Yolanda's Research," *Spirits Tincture*, Nov. 2016.

www.ingramcontent.com/pod-product-compliance
Lightning Source LLC
Chambersburg PA
CBHW071359230926
51846CB00002B/596

* 9 7 8 9 1 9 8 9 5 9 9 3 2 *